Table of Contents

[INTRODUCTION ....2](#)

[PART 1 ....5](#)

[PART 2 ....8](#)

[PART 3 ....11](#)

[PART 4 ....14](#)

[PART 5 ....17](#)

[PART 6 ....20](#)

[PART 7 ....22](#)

PART 8..**25**

PART 9..**28**

PART 10..**31**

PART 11..**33**

CONCLUSION..**35**

Part 1

1.
Why are ions attracted to one another?
Ions are attracted to one another because of their charges.

2.
What are ions with positive charges?
Ions with positive charges are cations.

3.
What are ions with negative charges?
Ions with negative charges are anions.

4.
The breaking of ionic bonds is called?
Ionization or dissociation.

5.
What is the pH of a cell?
The pH of a cell is 6.8

6.
Why is water cohesive?
Water is cohesive because of its hydrogen bonds.

7.
What kind of bonds do carbon atoms form?
Carbon atoms form covalent bonds.

8.

Are covalent bonds weakened in an aqueous solution?

No, covalent bonds are not weakened in an aqueous solution.

9.

What is energy?

Energy is the ability to do work.

10.

What is matter?

Matter is anything that has mass and occupies space.

11.

What are compounds?

Compounds are chemical substances of two or more elements.

12.

What are elements?

Elements are pure substances that cannot be broken down or decomposed into two or more substances.

13.

Why is an atom neutral?

An atom is neutral because the number of protons cancels the number of electrons.

14.
Is the number of protons different for each element?
No, it is unique for each element.

Part 2

1.

What determines the atomic number?
The number of protons.

2.

Which is stronger, covalent or ionic bond?
Covalent bonds.

3.

What is the body's normal pH?
The body's normal pH is 7.35 to 7.45

4.

What is the difference between organic and inorganic compounds?
Organic compounds contain carbon.

5.

Give examples of inorganic compounds in the body.
Examples of inorganic compounds in the body are oxygen, water, and acid and bases.

6.

Give examples of organic compounds in the body.
Examples of organic compounds in the body are carbohydrates, nucleic acids, lipids, and proteins.

7.

What are the basic protein units?

Amino acids are the basic protein units.

8.

What are the two most important chemicals in the body?

The two most important chemicals in the body are acid and base.

9.

What is acidosis?

Acidosis is when the pH of the blood is less than 7.35, which is acidic.

10.

What is alkalosis?

Alkalosis is when the pH of the blood is above 7.45

11.

What is proteome?

Proteome consists of all the proteins produced by the body.

12.

How many amino acids exist?

There are 20 amino acids.

13.

Name the groups of amino acids.

An amine group and an organic acid group.

14.

What are heavy metals?

Heavy metals are metals toxic to the body.

15.

What is ketosis?

Ketosis is a form of acidosis due to the excess of ketone bodies produced by the degradation of fats in the blood.

Part 3

1.

Can enzymes cause chemical reactions?

No, enzymes cannot cause chemical reactions.

2.

Enzymes are what type of proteins?

Enzymes are globular proteins.

3.

What factors can damage proteins?

pH and excessive temperature can damage proteins.

4.

What are denatured proteins?

Denatured proteins are when proteins lose their three-dimensional shape.

5.

What makes each of the amino acids unique?

The R group.

6.

Why can amino acids act as a base or acid?

Amino acids can act as a base or acid due to their amino acid and organic group.

7.

What are amphipathic molecules?

Amphipathic molecules are molecules that possess both polar and non-polar regions.

8.

What are trans fat?

Trans fat are oils that have been solidified by the addition of hydrogen atoms.

9.

What are unsaturated fats?

Unsaturated fats are fatty acids in which the carbon chain contains one or more double bonds.

10.

What are saturated fats?

Saturated fats are chains of fatty acids, which contain only simple covalent bonds between their carbon atoms.

11.

Why do women withstand cold better than men?

Women can withstand cold better than men because they have more fat in their hypodermis.

12.

Where are triglycerides?

In the hypodermis.

13.

Why is fat the most effective way for the body to concentrate and store usable energy?

This is because fat does not mix with water and when oxidized, it produces large amounts of energy.

14.

Why are fatty acids non-polar?

Fatty acids are non-polar because they are formed of hydrocarbon chains.

15.

What are glycerides composed of?

Glycerides are composed of glycerol and fatty acids.

Part 4

1.

Do a comparison between DNA and RNA.

DNA:

-Location: in the nucleus; Type of sugar: deoxyribose; Nitrogen bases: ACGT; Structure: double stranded.

RNA:

-Location: in the cytoplasm; Type of sugar: ribose; Nitrogen bases: ACGU; Structure: simple stranded.

2.

Name 2 functions of oxidases.

Oxidases remove hydrogen and add oxygen.

3.

What adds a molecule of water during hydrolysis?

Hydrolases.

4.

What is the main function of carbohydrates?

The main function of carbohydrates is energy production.

5.

What are the 2 polysaccharides essential to the body?

Glycogen and starch.

6.

What are weak acids?

Weak acids are acids that do not dissociate completely in water.

7.

Give an example of a strong acid.

HCl is an example of a strong acid.

8.

What are strong acids?

Strong acids are acids that do not dissociate completely in water.

9.

Give an example of a neutralization equation.

An example of a neutralization equation is H C l and NaOH which gives NaCl and H2O

10.

What are the products of a neutralization equation?

Salt and water

11.

What is neutralization?

Neutralization is a mixture of an acid and a base to produce water and salt.

12.

What represents the hooks in pH?

The hooks represent the concentration of a substance.

13.

What is the pH of a solution?

The pH of a solution is the measure of the acidity of a solution.

14.

What is a base?

A base is a proton acceptor.

15.

What is an acid?

An acid is a proton donor.

Part 5

1.

What are strong bases?

Strong bases are bases that dissociate completely in water and rapidly capture H+

2.

Give examples of weak acids.

Examples of weak acids are acetic acid and carbonic acid.

3.

What are buffers?

Buffers regulate pH.

4.

Are acids and bases electrolytes? And why?

Yes, acids and bases are electrolytes because they become ionized and dissociate in water, so they can conduct electric current.

5.

What is the most abundant salt in the body?

The most abundant salt in the body is calcium phosphate.

6.

What are electrolytes?

Electrolytes are substances that conduct electricity.

7.

What is a hydrolysis reaction?

A hydrolysis reaction is rupture under the action of water.

8.

The property of water that allows ionic compounds and other small reactive molecules dissociate in water is?

Its polarity.

9.

What is the advantage of the high thermal capacity of water?

The high thermal capacity of water regulates the temperature from sudden changes.

10.

Name the properties of water that make it a vital liquid.

Its polarity, its heat of vaporization, its reactivity, and its high thermal capacity.

11.

Organic compounds consist of molecules formed by what kind of bonds?

Organic compounds consist of molecules formed by covalent bonds.

12.

What do organic compounds contain?

Organic compounds contain carbon.

13.

Do biological catalysts become part of the products?

No, biological catalysts do not become part of the products.

14.

What are catalysts?

Catalysts are substances that increase the rate of chemical reactions.

15.

Identify some factors that may favour the collisions between particles and the course of reactions.

Catalysts, concentration, temperature, and particle size.

Part 6

1.

What does it mean that the living organism is an open system?

It means that it is a system that exchanges energy with its environment.

2.

Why does the chemistry of living organisms depend heavily on carbon?

This is because carbon is electro neutral and can also establish four covalent bonds with other carbon atoms or with other elements.

3.

Give some examples of polymers.

Proteins and carbohydrates are examples of polymers.

4.

What are polymers?

Polymers are molecules formed from monomers.

5.

Monomers are joined by?

Monomers are joined by a synthesis reaction by dehydration.

6.

What are the basic units of most carbohydrates?

The basic units of most carbohydrates are monosaccharides.

7.

What is the relationship between the size of a carbohydrate molecule and its solubility?

The larger the carbohydrate molecule, the less soluble it is in water.

8.

How are monosaccharides formed?

Monosaccharides are formed of a single chain containing carbon atoms, from 3 to 7 atoms.

9.

What do monosaccharides contain?

Monosaccharides contain carbon, hydrogen, and oxygen.

Part 7

1.

What is a dynamic equilibrium?

A dynamic equilibrium means that there is still formation and dissociation of molecules and products.

2.

Give an important condition for a reaction to have a chemical equilibrium.

For a reaction to have a chemical equilibrium, the system must be closed.

3.

What is an endothermic reaction?

An endothermic reaction is a reaction that absorbs energy.

4.

Give an example of exothermic reactions.

Catabolic reactions are examples of exothermic reactions.

5.

Give another name for an endothermic reaction.

An endergonic reaction.

6.

Give another name for an exothermic reaction.

An exergonic reaction.

7.

What does reduced in an oxidation-reduction reaction mean?

It is the reagent that accepts electrons.

8.

What does oxidized in an oxidation-reduction reaction mean?

It is the reagent that loses electrons.

9.

What is an exchange reaction?

An exchange reaction involves synthesis and degradation.

10.

Give an example of a degradation reaction.

AB results to A + B

11.

What is a degradation reaction?

A degradation reaction is when a molecule is shredded into smaller molecules.

12.

Give an example of a synthesis reaction.

A + B results to AB

13.

What is a synthesis reaction?

A synthesis reaction is when atoms or molecules combine to form a larger; a more complex molecule.

14.

What are the 3 modes of most chemical reactions?

The three modes of most chemical reactions are synthesis, degradation, or exchange.

15.

Name 3 types of chemical bonds.

Three types of chemical bonds are covalent, ionic, and hydrogen bonds.

Part 8

1.

What does each electron layer of an atom represent?

A level of energy.

2.

Why are the electrons of the inner layers generally not involved in the bonds?

Electrons of the inner layers are not involved in the bonds due to the strong attraction exerted by the nucleus.

3.

What are anions?

Anions are electron acceptors and they carry a negative charge.

4.

What are cations?

Cations are electron givers and they carry a positive charge.

5.

When 2 atoms share a pair of electrons, what kind of bonds do they form?

They form a simple covalent bond.

6.

When atoms share 2 or 3 pair of electrons what kind of bonds do they form?

They form double or triple covalent bonds.

7.

What are non-polar molecules?

Non-polar molecules have poles that are electrically neutral.

8.

What are polar molecules?

Polar molecules have unequal pairs of electrons in their poles.

9.

Give examples of electronegative atoms.

Oxygen and chlorine are examples of electronegative atoms.

10.

What are electronegative atoms?

Electronegative atoms are atoms that attract electrons.

11.

What are electropositive atoms?

Electropositive atoms are atoms that lose their own valence electrons.

12.

Give examples of electropositive atoms.

Sodium and potassium are examples of electropositive atoms.

Part 9

1.
What is a solution?
A solution consists of homogeneous mixtures of substances.

2.
What are the states of a solution?
The states of a solution are solid, liquid, or gas.

3.
What is a solvent?
A solvent is the most abundant substance in a solution.

4.
What is a solute?
A solute is the least abundant substance in a solution.

5.
What is the main solvent of the body?
The main solvent of the body is water.

6.
How do we describe true solutions?
We describe true solutions by indicating their concentration.

7.

How do we express the concentration of a solution?

We express the concentration of a solution by their molarity, milligrams, or percentage.

8.

Why is the mole as a unit of measure in the preparation of a solution important?

It is important because of its precision.

9.

Colloids are also called?

Colloids are also called emulsions.

10.

What are suspensions?

Suspensions are heterogeneous mixtures containing large particles, often visible, which tend to settle down.

11.

Give an example of a suspension.

Blood is an example of a suspension.

12.

What is the difference between mixtures and compounds?

Mixtures are physically compounded and compounds are chemically compounded.

13.

Are all mixtures homogeneous?

No, not all mixtures are homogeneous.

14.

Are all compounds heterogeneous?

No. Compounds are always homogenous.

Part 10

1.
What is biochemistry?
Biochemistry is the chemistry of living matter.

2.
What is chemistry?
Chemistry is the study of the nature of matter.

3.
What is the difference between mass and weight?
Mass is constant but weight depends on gravity.

4.
Does energy contain mass and volume?
No, energy does not contain mass and volume.

5.
What is an atom?
An atom is the unit of matter.

6.
What are the particles of an atom?
The particles of an atom are protons, neutrons, and electrons.

7.
What is in the centre of an atom?
The centre of an atom is the nucleus.

8.

What is the charge of a proton?

The charge of a proton is positive.

9.

Neutrons carry what charge?

Neutrons carry no charge.

10.

What charges do electrons carry?

Electrons carry negative charges.

11.

Why do elements have different properties?

Elements have different properties because elements are composed of a different number of protons, neutrons, and electrons.

12.

What determines the chemical behaviour of atoms?

The chemical behaviour of atoms is determined by their electrons.

13.

What are radioisotopes?

Radioisotopes are isotopes that are part of an atomic disintegration.

Part 11

1.

Explain kinetic energy.

Kinetic energy is energy represented by movement.

2.

Explain potential energy.

Potential energy is energy in a stored form.

3.

Explain chemical energy in the body.

Chemical energy is energy stored as ATP.

4.

Explain electrical energy in the body.

Electrical energy is the energy of the nervous system through electrical impulses.

5.

Explain mechanical energy in the body.

Mechanical energy is the energy that directly produces a movement of matter.

6.

Explain radiation energy.

Radiation energy is an electromagnetic energy that propagates in the form of waves.

7.

Give some examples of elements.

Examples of elements include oxygen, carbon, and hydrogen.

8.

Why do we not add the mass of the electrons to the mass number?

This is because it is negligible.

9.

What are isotopes?

Isotopes are structures that have the same number of protons but a different number of neutrons.

10.

What is radioactivity?

Radioactivity is the process of atomic disintegration.

Conclusion

Thank you once again for purchasing this book. I hope it has helped you in your journey to understanding the anatomy and physiology of the human body.

Please, if you enjoyed this book, I would like you to leave a review. It'd be appreciated.

Thank you.

www.ingramcontent.com/pod-product-compliance
Lightning Source LLC
Chambersburg PA
CBHW031558210526
45464CB00003B/1335

ANATOMY AND PHYSIOLOGY CHEMISTRY AND THE BODY

THINGS YOU SHOULD KNOW
(QUESTIONS AND ANSWERS)

By Rumi Michael Leigh

Introduction

I would like to thank and congratulate you for purchasing this book, "*Anatomy and physiology, chemistry and the body, things you should know (questions and answers)*" series.

This book will help you understand, revise and have a good general knowledge and keywords of the human anatomy and physiology.

Thanks again for purchasing this book, I hope you enjoy it!